I0797654

Aquarium

Julie Murray

Abdo Kids Junior
is an Imprint of Abdo Kids
abdobooks.com

Abdo
FIELD TRIPS
Kids

abdobooks.com

Published by Abdo Kids, a division of ABDO, P.O. Box 398166, Minneapolis, Minnesota 55439.

Abdo Kids Junior™ is a trademark and logo of Abdo Kids.

Printed in the United States of America, North Mankato, Minnesota.

102019

012020

Photo Credits: iStock, Shutterstock

Production Contributors: Teddy Borth, Jennie Forsberg, Grace Hansen

Design Contributors: Christina Doffing, Candice Keimig, Dorothy Toth

Library of Congress Control Number: 2019941196

Publisher's Cataloging-in-Publication Data

Names: Murray, Julie, author.

Title: Aquarium / by Julie Murray

Description: Minneapolis, Minnesota : Abdo Kids, 2020 | Series: Field trips | Includes online resources and index.

Identifiers: ISBN 9781532188718 (lib. bdg.) | ISBN 9781532189203 (ebook) | ISBN 9781098200183 (Read-to-Me ebook)

Subjects: LCSH: Public aquariums--Juvenile literature. | Aquatic animals--Juvenile literature. | Aquariums--Juvenile literature. | School field trips--Juvenile literature.

Classification: DDC 371.384--dc23

Table of Contents

Aquarium 4

What is at
an Aquarium? 22

Glossary. 23

Index 24

Abdo Kids Code. 24

Aquarium

It's field trip day! The class is going to the aquarium.

There are lots of fish. Some are big. Some are small.

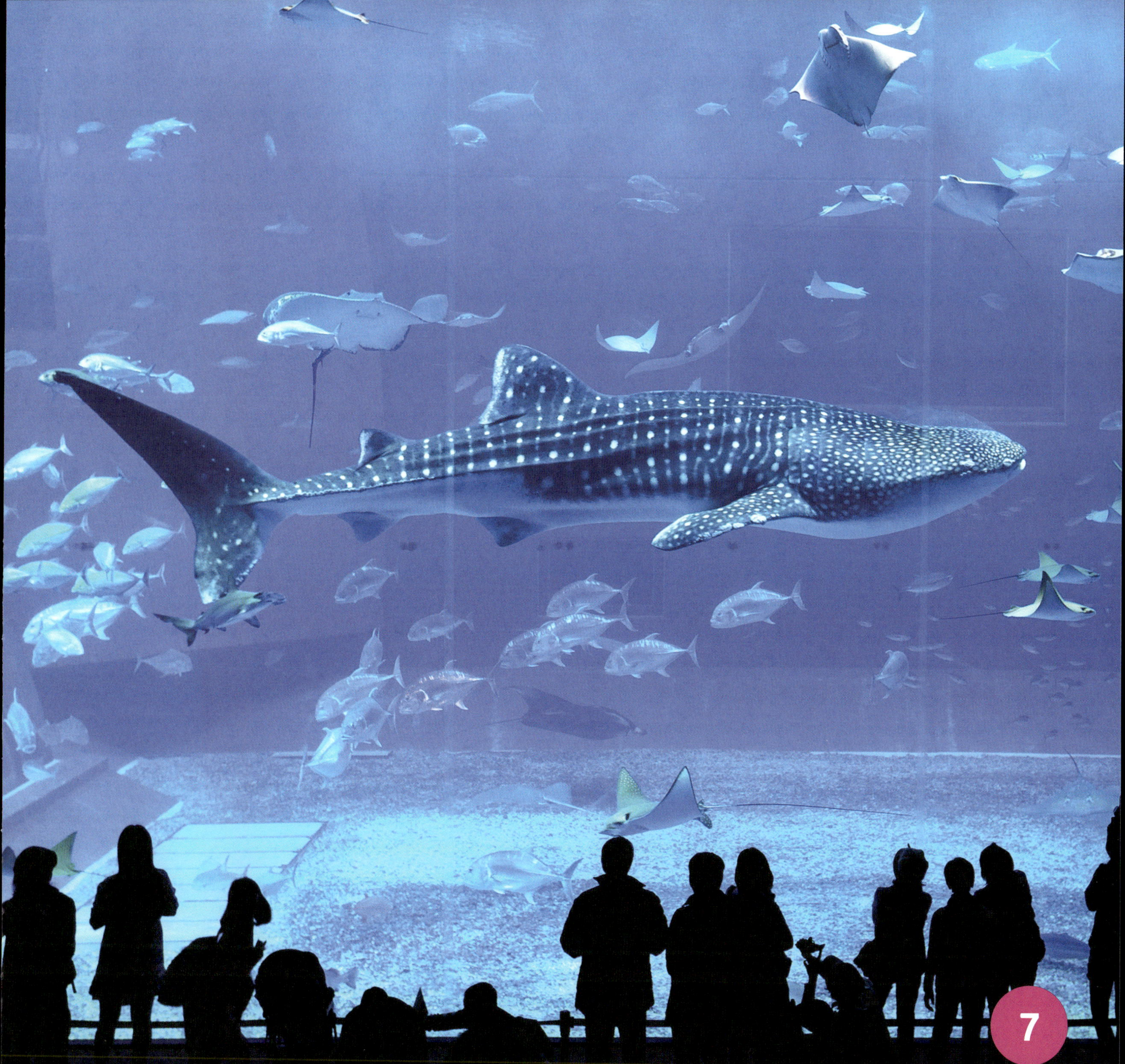

Ted finds a sea turtle.

It has **flippers**.

Jose touches the ray.

It is smooth.

Liz sees a **coral reef**.

It has lots of colors.

Hun sees the penguins.

They are swimming.

The sharks are big.

They swim slowly.

The worker feeds the fish.

Mary watches.

Have you been to
an aquarium?

What is at an Aquarium?

fish

penguins

sharks

turtles

Glossary

coral reef
a mound or ridge of coral skeletons and other materials that forms in warm, shallow sea waters.

flipper
a wide, flat limb on a turtle that is used for swimming.

Index

coral reef 12

feeding 18

fish 6, 18

penguins 14

sea ray 10

sea turtle 8

shark 16

workers 18

Visit **abdokids.com**

to access crafts, games,

videos, and more!

Use Abdo Kids code

FAK8718

or scan this QR code!